Tops'l Books
9 Queen Victoria Street, Reading, England

PR
and
the Brixham smacks

by John Corin

With old photographs from the great days of Devon's sailing trawlers, and of Provident, one of the few survivors of the era.

Acknowledgements

The author is indebted to many people in Brixham and elsewhere for their help, in particular Mr. Ronald Pillar, Mr. Larry Bubeer, Mr. John Horsley, Mr. Tony Pawlyn and Mr. Clifton Pender.

Vice-Admiral Sir Patrick Bayly, Director of the Maritime Trust, kindly read the manuscript and made a number of suggestions. Indeed, the writing of the book was his idea. The typing of the manuscript was kindly carried out by Mrs. D. H. Taylor.

Anyone writing on the sailing trawlers owes a very great debt to the late Edgar J. March and his study of the subject.

The author is also indebted to the various Island Cruising Club Skippers who have endeavoured to teach him something of the art of sailing *Provident*.

A further word of thanks is due to Captain S. T. Smith, Editor of Olsen's Nautical Almanack for help in tracing details of many old smacks.

Cover picture

- From an oil painting of *Provident* by the marine artist Peter Stuckey, a life long student of working sail. Like the author he is also a member of the Island Cruising Club and devotee of *Provident*. The scene records an incident in the Channel when *Provident* had been obliged in a fresh breeze, to alter course 20° to windward to avoid a steamer. A sea struck the port quarter, washing one of the crew into the lee scuppers and invading the doghouse. The author was at the wheel at the time and provided the artist with a word picture of the incident which he has faithfully interpreted. The painting in the possession of the author, is reproduced by kind permission of Peter Stuckey.

Title page

- *Provident* in quiet weather, topsails, working staysail and No. 1 jib set, approaching her present home port, Salcombe.

Photo: Ted Pearce.

Opposite

- The breakwater at Brixham in 1868 in the early days of the fishing boom. By that date the breakwater had only reached about a third of its final length of 3,000 feet, which took until 1921 to complete. By that time the days of the big ketches were numbered.

Photo: D. F. & H. M. Stone, Brixham.

Contents

Forewords

**by John A. Baylay, T.D.,
Founder-President,
Island Cruising Club.**

Just an ordinary Brixham trawler.
The war – and *Provident* lying off Helford as a Navy store ship.
A Brigadier in B.A.O.R. wanting such a trawler as a yacht in Hamburg.
A very seasick honeymoon ending in Salcombe, resulting in an unusable asset.
An embryo Club wanting a cruising yacht big enough for a large crew of members.
A Maritime Trust wishing to preserve the best of our sea heritage.
This strange amalgam has resulted in the preservation of an important example of 'working sail'.
Between 1951 and 1980 nearly 100,000 miles of sailing.
The enjoyment of the sea and training in seamanship of several thousands of men and women.
A Club flagship of renown and repute in press, television, film and in heaven knows how many harbours and clubhouses.

May I be forgiven in misquoting – "Tall Ships from little acorns grow".

Publisher's note
The author has assigned the royalties in this book to be shared equally by the Maritime Trust and the Island Cruising Club to further the excellent work of these two bodies. The contributors of the two Forewords have, as readers will observe, strikingly similar, but different names. There is no printer's error. To add further to the coincidence the Royal Navy Squadron which opens the story in Chapter One, was commanded by another Admiral Bayly.

**by Vice-Admiral Sir Patrick Bayly, KBE, CB, DSC,
Director of The Maritime Trust.**

In 1934 Frank Carr identified over 200 distinct types of coastal craft around these islands, the majority being fishing boats, from deep sea trawlers to beach boats. To the lay observer they were all 'fishing smacks' but usually they were built locally, for local people, to suit local conditions and to designs that had evolved perhaps over several centuries. Each could be placed by her prominent fishing registration number with its self-evident port letters. (Even today there are 140 such ports.)
One of the most widely known types was the Brixham trawler. Sailing from the harbours of the south west, often within easy view of the landsmen, the trawlers added a picturesque dimension to the scene which could hardly be matched elsewhere. Their sturdy and seaworthy hulls, topped by sails the colour of the Devonian earth, were a sight which could stir the least romantic soul.
Now they have gone, and the other 200 types with them, but it is fitting that the life of one, the *Provident*, should be chronicled while there are still a few witnesses to tell the tale.
The Maritime Trust is proud to own the *Provident* and welcomes this fine account of her life and the world in which she sailed.

● *Provident* close-hauled, with the mizzen topsail not set. The crew are eyeing *Hoshi*, a vintage Edwardian schooner, built by Camper & Nicholson in 1908, as a gentleman's yacht. She is also in the Island Cruising Club fleet.
Photo: Roy J. Westlake, Plymouth.

Provident's Prologue

ON THE last day of the year 1914 the Fifth Battle Squadron, with *H.M.S. Lord Nelson* as flagship, was cruising in the Channel before going to gunnery practice. On board one of the ships was F. E. Little, of the Royal Marine Light Infantry, and in the early hours of New Year's Day, 1915, they were not far, in terms of miles, from his home in Kingswear. It was a clear but cloudy night with a calm sea, when, cruising at 10 knots in line ahead, the squadron was in a position some 12 miles south-east of Berry Head, Brixham. Eighth and last ship in the line was the battleship *Formidable* and at 02.20 a torpedo struck her on the port side with a terrible explosion.

Almost immediately the whole squadron did a 180° turn. Remembering, many years later what happened next Mr. Little wrote: "*Diamond* and *Topaze* were detailed to stand by and render assistance. The stricken ship started to list to port and settle by the bows, when suddenly the wind started to increase and rapidly became gale force, the moon disappeared and rescue work became most difficult. The sea very soon became mountainous and boats, rafts and so forth, which had contained survivors were swept clean. The two cruisers continued their efforts at rescue, now exceeedingly difficult, as the sea was full of men endeavouring to swim or cling to some floating jetsam and, inevitably, many were drowned.

"The trawler *Provident* arrived on the scene and set about rescue work, even though she must have been having an appalling time in that sort of sea, but nevertheless she was able to grab about 70 men from the waves, while the cruisers got about another 80 between them. This was most regrettably the total of survivors from a crew of 750.

"In the terrible seas that were running after *Formidable* sank one cannot but marvel how a small ship like *Provident* could have even ridden them."

Other accounts record that *Provident* was running for her home port of Brixham when she sighted *Formidable's* ship's launch. An oil painting in the possession of Mr. Ronald Pillar, a nephew of *Provident's* Skipper, William Pillar, shows the launch, with a shirt fluttering from an up-raised oar, and the trawler pitching into the 30 feet waves under a reefed mainsail.

Under the prevailing conditions of wind and sea getting alongside the launch must have been a daunting task for the 32-year-old skipper and his crew, even with their years of experience of trawling, involving handling the vessel in all conditions. To get to leeward of the launch, in order to make as controlled an approach as possible, they were obliged to gybe three times, a sailing manoeuvre, which in a heavily sparred vessel in gale conditions is not lightly undertaken. King George V himself commented on this when he received the crew at Buckingham Palace later in the year, something which he was qualified to do, having not only had sail training in his Naval career, but experience in sailing his famous J Class racing yacht *Britannia.*

● The first *Provident, BM291* in which Skipper William Pillar rescued 72 survivors from the torpedoed battleship H.M.S. *Formidable* in a gale on New Year's Day 1915. *Provident* was hove to on the starboard tack under reefed mainsail, off Start Point when a cutter from the *Formidable,* flying a sailor's scarf from an upraised oar, was sighted by John Clarke, the second hand.

Photo: *Illustrated London News.*

However, after the third gybe *Provident* was able to bring the launch under the lee quarter and begin the task of getting more than 70 men, some of them wounded, over the side into safety on board. Fortunately a trawler has a relatively low freeboard aft to facilitate hauling the trawl on board, but even so it must have been difficult to get the survivors on board with a heavy sea running.

The survivors were landed at Brixham and, at the instigation of Winston Churchill, the Admiralty made an award of £250 to William Pillar in recognition of his efforts and £100 to each of the crew, William Carter and John Clarke. Daniel Taylor, the boy cook received £50, and Leonard Pillar, the Skipper's nephew, £25. The crew also received the silver gallantry medal for saving life at sea.

The *Provident* of this story, *BM291,* was not the Brixham trawler we see today, for on November 28th, 1916, she was stopped by a U-boat, the crew given time to leave her, and then she was sent to the bottom. But in this melancholy end there lies the genesis of our story. A new *Provident* was built, registered as *BM28,* and she survives today as a reminder of that gallant rescue, and of the great fleet of sailing smacks which once fished from the Devon port.

● Above, Skipper William Pillar as he appeared in a photograph in the *Illustrated London News* shortly after the rescue in 1915. He was then 32.

● The top photograph was taken some years before the outbreak of war, perhaps on board the first *Provident* when she was in one of the Cornish ports. Pillar is the centre one of the group of three men. Skipper Pillar survived in Brixham to the grand age of 87. The photograph below was taken in 1969, a year or so before his death, and shows him with one of the H.M.S. *Formidable* survivors.

Photos: National Maritime Museum.
S. G. Pine.

The Growth Of A Port

BRIXHAM can claim a large stake in the founding of the national trawling industry by developing the finest sailing fishing vessel of its kind in the world, and by colonising certain East Coast ports on which the industry has been chiefly based in modern times.

What manner of place is it, which built trawlers like *Provident*, and sent away families around the coast, to found fishing dynastics, whose descendants are to be found trawling the Distant Waters north of Norway and Russia, into the White Sea?

The little town's geographical advantages are neither numerous, nor generous, except in one respect. Brixham lies in the southern corner of Torbay, it looks towards the north, and is admirably protected from the prevailing wind, and the gales, which come from the south-west. To the eastward Berry Head runs out for about a mile. The Admiralty's Channel Pilot describes Torbay as being "sheltered in all winds from north-east through north and west, to south-by-west" and even, at the Brixham end, to a wind from the south-south-east.

In the Napoleonic Wars the Royal Navy was very well aware of the shelter against westerly gales afforded by Torbay. If the Western Squadron blockading Brest were forced to leave by a westerly gale they could run for Torbay, secure in the knowledge that if they could not stay, the French could not beat out of Brest.

Torbay has the considerable advantage that sailing vessels can enter and leave it in almost any wind direction. Thus fishing vessels would rarely be either wind-bound in port, or unable to enter to land their catches.

Brixham was originally a small inlet in the corner of the Bay and nature had not provided a natural harbour on the scale of that at neighbouring Dartmouth, or to compare with other West Country ports like Salcombe, Plymouth, Fowey or Falmouth. However, by the middle of the 18th century the fishing industry had acquired sufficient importance for a move to be made to improve the limited existing facilities of the original inner harbour, which consisted of nothing much more than quay walls and a short jetty. Twelve Brixham men and their wives joined together and bought the harbour from the Lord of the Manor in 1759, becoming the 'Quay Lords and Ladies', a title which was still in use when the present *Provident* came into service in 1924.

- Below, *Gleaner BM303* coming into harbour at Brixham. She was 25 tons, built in the port in 1890 for W. H. Nowell. The great beam of the trawl can be seen along the port quarter with the 'iron' projecting beyond the stern. Above, remnants of the fleet putting to sea in the twilight days of sail. The centre vessel is *Sanspariel BM326* built in 1912 for W. R. Edhouse.

Photos: *The Times* Picture Library.
L. S. Lyddon.

BM-303
BM-303

Soon after 1795 the Eastern Quay was built and in 1799 an Act was obtained to build the New Pier. Because of Torbay's importance to the Navy there were also improved watering and victualling facilities provided.

About the time the Eastern Quay was built Brixham seems to have given up the North Sea drift net herring fishery, which had been a staple, and turned more to trawling. By 1780 some trawlers had started to go as far as the Yorkshire coast and land fish there. Despite its geographical remoteness the advent of the much improved turnpike roads had made Brixham the largest wholesale fish market in the West of England, able to supply Exeter, Bath and Salisbury. By the early years of the 19th century Brixham trawlermen were working not only the English Channel, despite the hazards from French privateers, but had found their way to the Bristol Channel, the Irish Sea and north to the Isle of Man grounds. In 1818 Dublin Bay was trawled for the first time. By 1833 official records showed 112 fishing vessels in the port, most of which were going to the North Sea. A decade before there had been only 89 decked boats. By 1850 Brixham had 210 trawlers, of which 130 were between 30 and 50 tons, and the Port's great colonisation era was well under way.

In the provincial England of the 19th century people settled in their occupations did not normally move far out of their own chimney smoke. Their work was within walking distance. Farmers, tradesmen and professional folk mostly lived and worked in the same place for generations. The fisherman was an exception. He was a hunter who sought fish on a seasonal basis wherever he could find them, the movement being partly dictated by the sea conditions of winter or summer and partly by fish migrations. It was not unknown for a West Country fisherman to circumnavigate the island of Britain, moving from one ground to another in the course of the seasons of a year. This was done by men without formal training in navigation and pilotage, with no navigational aids on board, beyond a lead line and a magnetic compass of doubtful deviation. They ventured forth, as their forefathers had done, learning by the hallowed, and sometimes disastrous practice of 'trial and error'. They acquired the necessary local knowledge, and they operated without charts. "Charts", said the old time fishermen, "are alright for them as knows the locality!"

In practice they usually knew very well where they were. They made much use of the lead line to find what depth of water they were in. The lead also, when armed with a piece of tallow, brought up a sample of the sea bottom which gave the old time fishermen another indication of his position; sometimes, it is said, by actually tasting the sample!

Apart from that he developed a subconscious constant sense of his whereabouts. Professional navigators 'doing it by the book' in coastal waters in the Second World War, found that fishermen shipmates could give them a rough position apparently without doing any navigational work at all (see "Escort" by D. A. Rayner, Futura 1975 p. 30).

It is therefore no surprise to find that by 1833 as many as 60 to 70 Brixham smacks were going to Ramsgate for the winter to trawl from there up to Christmas. Some stayed on until the summer, but by June it was becoming too warm for their fish to reach the London market in prime condition. In the summer of that same year 12 smacks went to the Bristol Channel and six to Liverpool.

- *Encourage BM63* (right) 54 tons, built in 1926, was the last sailing smack from R. Jackman's new yard and built for H. Lang of Brixham. Her length was 69 feet and beam 19 feet. Brixham men seem to have had a penchant for naming their vessels in a rather unusual manner, using words like Encourage, Inspire, Vigilance, Prevalent, Provident, Replete, etc. *William & Sam BM352,* 25 tons (above) was built by J. W. & A. Upham in 1917 for T. Jackman. She won the mule class in the 1928 Regatta. Her hull was light blue and she was converted to a yacht in 1930, but broke her back by accident when being put on the blocks and was broken up at Galmpton on the River Dart in 1950. She is setting the big overlapping 'towing' staysail used by the fishermen in favourable conditions of wind to urge an extra knot or two out of their smacks. At one stage in her career she was re-named *Red Ensign.*

In 1837 there came a break through in North Sea fishing—the discovery of the Silver Pits. As with most major discoveries there is more than one claimant for the honour of having first made it, but according to one authority it was the Brixham man James Stubbs, who had taken his small trawler to Hull in 1832. He discovered the Silver Pits by chance when driven off course in a gale with the trawl down and to find them again consulted the charts of the colliers lying in Scarborough Harbour. He, of course, had none.

The Silver Pits, south of the Dogger Bank, are a deep depression where in cold winters the sole congregate in large numbers. On one occasion Stubbs had 2,040 pairs of soles in the trawl in one haul. The discovery must have undoubtedly helped the foundation of the trawling industry by the Brixham colony in Hull. By the middle of the next decade there were 21 trawlers in that Port and the industry grew, turned to steam trawlers and eventually reached out to the waters off Iceland and north of the Arctic Circle. Even as late as 1893 the accents of Devon were to be heard on Humberside.

In 1850 Brixham had acquired a total of 210 trawlers. During the next decade many of them moved to Hull, Grimsby, Lowestoft, and Scarborough, while their rivals from Barking (now lost in the London sprawl, but once a premier fishing port) moved in on Harwich and Lowestoft. By 1860 there was very little of the North Sea which had not seen the tanned sails of the Devon trawlers. On the other side of Britain a colony of Brixham men went to Tenby.

Another important factor was the coming of the railway, which reached Penzance in 1859, opening up opportunities for Brixham men to trawl and land fish as far west as Penzance and Newlyn to be put on the train. Their arrival was not popular. The trawlers started going over the mackerel grounds between February and April and the Cornish drift net fishermen later gave evidence to a Government inquiry, in 1878, that they deliberately fouled and cut the drift nets. Not for the first, or the last time there was trouble and violence in an industry where interests and feelings can easily clash.

Along with the development of its fishing fleet the merchant service had also expanded remarkably in Brixham. In 1850 the Port had six brigs of 170 tons and 140 schooners of 60 to 180 tons, many of them trading overseas. The harbour facilities remained quite inadequate. If anything they were diminished by extending the town over the innermost part of the harbour. Outside the Inner Harbour, as it is now, craft were obliged to lie on individual moorings exposed to easterly gales. It would seem however that an under-set of current to windward, which increased with the wind, made these moorings more tenable than might at first appear. In 1836 a plan was drawn up for a great breakwater starting from Mackerel Rock, to the east of the town and extending in a north westerly direction. However, while the initiative to build individual ships and trawlers was there, the administrative machinery and enthusiasm for a community enterprise were sadly lacking. In 1843 a little over £300 had been raised, but the wealthier shipowners were apathetic. The same situation seems to have occurred in Salcombe, *Provident's* base port since 1950. There is a fine natural harbour and Salcombe was famous for its schooners, but the facilities by way of jetties and quays, let alone desirable features like dry-docks and gridirons remained very limited.

Nevertheless in Brixham 1,000 feet of breakwater had been built by 1866. In that year a devastating demonstration of the necessity of such works occurred. One of those terrible gales, which occur perhaps once in a century in their concentration on one section of coast, struck Torbay. On the night of January 10th-11th over 100 ships were sheltering in the Bay when the wind backed to the north-east and mounted to hurricane force. Sixty vessels and at least 100 lives were lost. Thirty years later there was still 700 feet of breakwater left to build and the Local Authority took over the task from the Harbour Commissioners. By that time about £23,000 had been spent, much of it raised by the fishermen paying a levy of three pence in the £1 on all fish landed.

Not until 1921 was the breakwater at last completed to its present 3,000 feet length, but in the nineties shelter for 300 large and small vessels had been achieved, ten times the capacity of the inner harbour on neap tides.

Down west at Newlyn a new harbour had been built much more speedily and was of great benefit to the Brixham men who used this port extensively. Newlyn had the same problem as Brixham. There was shelter from the south-west, but its small quay and inner harbour were inadequate for the 400 mackerel and pilchard boats earning a living there by 1873. The visiting Brixham trawlers added to the overcrowding. On the initiative of the local Vicar a new large harbour of 40 acres was constructed, and completed by public subscription between 1884 and 1896.

As a good port for landing catches Newlyn became a home from home for the Brixham smacks, but although the 'Brickies' earned the respect of the local fishermen for their sea-keeping qualities, Brixham men never established a permanent base in Newlyn as they had done elsewhere. To those acquainted with Cornwall, and Newlyn in particular, this is scarcely a matter for surprise.

There is perhaps some justice in the fact that the Vicar who was behind the harbour improvement at Newlyn, which benefited Brixham, was a Devon man by birth, even if he did bear the name of Wladislaw Lach-Szyrma. Among many attributes he was a pioneer of science-fiction.

In the second half of the 19th century Brixham was not entirely without its own men of public spirited initiative, even if it was dilatory in the building of the breakwater. R. W. Wolston built two miles of broad gauge railway from Torquay to Brixham in 1868, virtually at his own expense, in order to serve the fishing port. It is true that the station was inconveniently above the town, but fish left it by train almost every day until it closed in 1963.

Nevertheless Brixham remained remote from supplies of cheap coal and the fishing fleet at Brixham did not go over to steam power, as did those ports which Brixham had colonised on the East Coast of England, where the proximity of the coalfields made steam a more economic proposition.

Brixham stayed with sail and in the years immediately before the First World War reached its zenith as a fishing port. In 1906 there were 220 first class smacks of 15 tons and over registered in Brixham and over 3,300 tons of fish landed in the town, apart from catches taken into other ports. Immediately before the outbreak of the war there were about 140 big ketches working.

During the war 35 smacks were sunk by U-boats. Six went down in one week and then an escorted fleet fishing system was instituted. *Provident's* predecessor of the same name was one of the victims as we have seen. After the war one might have expected the sailing trawlers to fade away or at least adopt some form of internal combustion engine, but, in fact, a new generation of trawlers was built, 15 of them over 40 tons. The present *Provident* was one of this generation. So also was *Valerian BM161,* of which there is a model in the Science Museum and also a fine model in the Conservative Club at Brixham, which was built by Mr. Ronald Pillar, a shipwright, and nephew of William Pillar. The last sailing trawlers to be built were *Servabo* and *Ruby Eileen* in 1927.

In Cornwall the internal combustion engine was universally adopted for fishing craft in the inter war years, but not by Brixham. It seems likely that there were no suitable engines available with sufficient power for a trawler. The Cornish boats, handling drift nets, long lines, or crab pots, only needed sufficient power to make a passage to and from the fishing grounds. They used either semi-diesels or petrol/paraffin engines. A trawler needs considerably more power for dragging the gear along the sea bed. Pure diesels with a suitable power weight ratio and an electric start, instead of air bottles, did not come along until after the Second World War.

In 1933 15 first class vessels, remained, three with some form of auxiliary power. The end was not far off and by 1938 most of the big ketches lay in Brixham Harbour awaiting buyers. That year, or the one before, must have been the last fishing season of the Brixham smacks. There must have been a 1937 season as the *Girl Inez* is recorded as being lost by fire while fishing off the Irish Coast in that year. At some time in those last years of peace the last sailing trawler on the fishing grounds must have hauled her old fashioned beam trawl for the last time and made back to harbour to land her catch. A great chapter in the history of Britain's maritime endeavour was closed.

It probably appeared to be the end of Brixham's history as a fishing port of any significance. Nothing could take away Brixham's record in the history of trawling, but after the Second World War the will to follow the old ways seemed to have ebbed away. By 1964 Brixham had reached its nadir, but the spirit of the place had not quite flickered out and was there to be re-kindled by a change of fortune. The new 12-mile international fishing limit was introduced and a whole class of Continental power trawlers, which fished in the English Channel, came on the market at knock down prices and were bought by Brixham men. A Fishermen's Co-operative for marketing fish came into being. In 1968 there were eight trawlers, in 1974 there were 56 and a new quay had been built out into the harbour, at a cost of £330,000, providing nine deep water berths, an ice plant, fish auction floor, offices and a heave-up slip.

Into this modern, thriving scene, comes from time to time each summer from 'The Port of the Past', *Provident*, slipping in past Berry Head, wearing the burgee of the Island Cruising Club and the flag of the Maritime Trust, and quietly taking up a mooring inside the breakwater, where so many hundreds of her predecessors have lain. Sometimes her entry is rather more boisterous, thrashing in from sea in Force 7-8 winds to sweep triumphantly through a convoy of modern coasters sheltering from the gale in the lee of the land, a living reminder of 'wooden ships and iron men'.

- Aloft on *Provident*. The photograph (left) of the forward side of the mainmast, shows the heel of the topmast, held in position by the trestle-tree and cross-trees. The topmast can be lowered by means of a heel rope and the removal of the fid which holds it up. In heavy winter gales it was the practice of the old sailing fishermen to house the topmast at sea, a drill not attempted by her modern amateur crews. The bottom photograph shows some of *Provident's* foredeck details. In her fishing days the bitts had the windlass immediately behind them and the pawl of the windlass was attached to the port bitt. As can be seen the bitts still fulfil their function of housing the heel of the bowsprit when it is in the run-out position. Smacksmen ran in the bowsprit as a means of reefing in very heavy weather.

Photos: Mike Brownlee and author.

- The launching of *Terminist BM321,* 37 tons in 1912. Families in best clothes are aboard for the occasion. Wide Edwardian ladies hats are much in evidence and excited children's faces look down over the bulwarks. *Terminist* was aptly named for she was the last vessel to be launched from Jackman's old Breakwater Yard at Brixham. The company moved alongside the rival Upham's yard where the last sailing smack, *Encourage BM63* seen on an earlier page, was built in 1926. In the same year Upham's built their last smack, *Vigilance BM76.* Both *Terminist* and *Vigilance* are still afloat. Opposite are the men who built *Terminist* and an earlier photograph, taken in 1880, of an unidentified smack being built at Jackman's Yard with planking half complete.

Photos: D. F. & H. M. Stone, Brixham.

A Smack At Work

THE BRIXHAM sailing trawler was evolved for the purpose of towing a trawl net, weighing one or two tons, over the bed of the sea, in open water far from her home port, in all kinds of weather. The demands of the trawl net as a fishing device dictated the characteristics of the vessel and it is important to understand what manner of contrivance it was and what those demands were.

Historically there are three principal kinds of net fishing around the British Isles, seining, drifting and trawling. The variations, ancient and modern, of these methods need not greatly concern us. Seining is, in its traditional form, a relatively shallow water, inshore method of fishing. A wall of net is shot around a shoal of fish, like pilchards or mullet. Then the net may be either drawn on to a beach, or drawn into a circle by the vessel, with its foot ropes resting on the bottom, and the fish scooped out.

In drift net fishing a wall of nets is shot, perhaps up to two miles in length. The fishing vessel, or drifter, rides to the nets into which the fish (herring, mackerel or pilchards) swim and become enmeshed. In due course the train of nets is hauled aboard the drifter and the fish shaken out of the net.

Seining called only for a large open boat, for which oars were an adequate form of propulsion. Seine boats were not, generally speaking, called upon to perform in other than fairly easy conditions of wind and sea. It is still a usual form of fishing in the Mediterranean Sea.

Drifters were obliged to go much further off shore to find the herring shoals, be able to make passages in all kinds of weather and ride safely to their nets in fairly arduous conditions. They did not however have to tow the nets in any way as a trawler did. A trawler was therefore a more powerful vessel whose crew had a different pattern of life with longer absences from home.

In this country the origins of trawling go back to the period of the Roman occupation. The Romans were very partial to oysters. The only practicable method of getting the oysters up from their undersea bed is by using a dredge, a kind of bagnet with its mouth held open by an iron frame. During the winter season small sailing and pulling craft can still be seen in Falmouth Harbour using precisely this method of fishing.

Although oyster beds, at least in the case of Falmouth, are not fish feeding grounds, it would be technically but a short step to start dredging for demersal fish. Demersal fish are those that feed on, or very close to, the bottom, such as cod and flatfish, like plaice, turbot and sole, as opposed to pelagic fish, which feed in the upper waters and include herring, pilchards, mackerel and mullet. The former are usually caught by trawling and the latter by drifting.

The first record of trawling in England is as far back as the reign of Edward III. Even then, in the years 1376-77, there was a 'conservation lobby' which presented a petition to Parliament and called for the prohibition of a 'subtlety contrived instrument called the wondyrchoum'! This consisted of a trawl net 18 feet long by 10 feet wide "of so small a mesh, no manner of fish, however small, entering within it can pass out and is compelled to remain therein and be taken by means of which instrument the fishermen aforesaid take so great

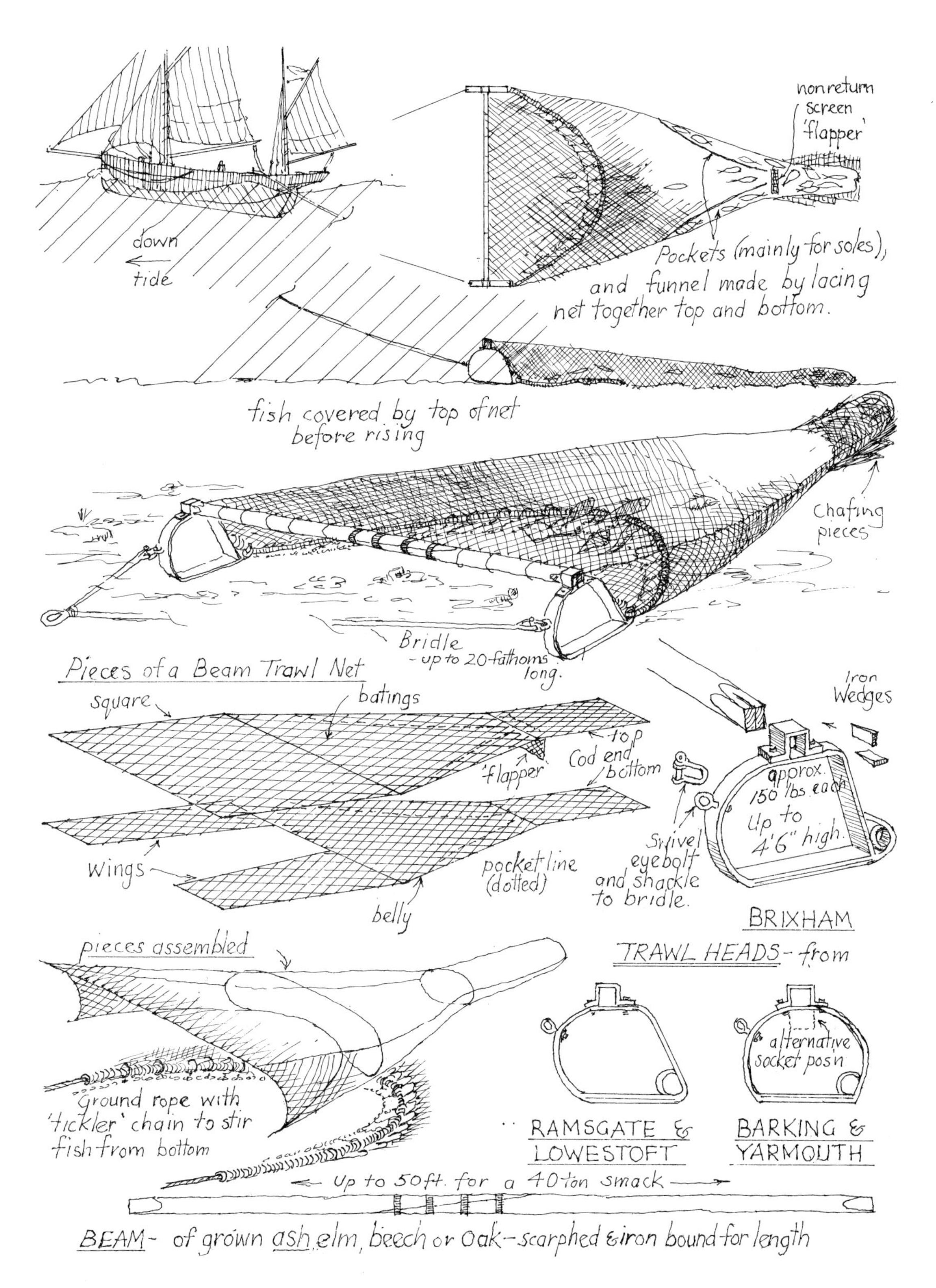

Drawings by Syd Brown

abundance of small fish aforesaid, that they know not what to do with them, but feed and fatten the pigs with them, to the great damage of the whole commons of the kingdom, and the destruction of the fisheries in like places, for which they pray remedy''.

Numerous government inquiries into the fishing industry, even in the 19th century when the harvest of the sea seemed abundant and endless, have heard evidence about the same problem and today it is still a major conservation worry. Indeed trawling has never been popular with fishermen of other persuasions. In Cornwall trawlers used to be referred to opprobriously as 'draggers'. Dislike of trawling may well have been an element in the celebrated Newlyn Riots of 1896, when troops and four Royal Navy destroyers were sent to restore order between the Newlyn men and visiting East Anglian crews.

One small mercy is that the very Chaucerian kind of word *wondyrchoum* went out of fashion to be replaced by the word trawl, of obscure origin, but possibly derived from the Middle Dutch word *traghelen,* 'to drag'. As well as changing name the net developed in size well beyond the dimensions of the 14th century wondyrchoum. By the latter part of the Victorian era it had become a large triangular bag net more like 100 feet long by 50 feet wide. Its shape can perhaps best be described as something like the old fashioned night cap which fitted over the head and then tapered away in a straggling conical fashion. Instead of a bobble the trawl finished in the narrow mesh cod-end, secured by a special knot. The term cod-end is not derived from the fact that it is where the fish accumulate; 'cod' is merely the Anglo-Saxon word for bag or purse.

The essential fitting for towing a trawl is something to keep the mouth of the net open. As the speed of a sailing vessel is so variable the only answer was the provision of a rigid beam. This was constructed of a strong, springy timber, more than one length being scarphed together, if necessary, and strengthened by iron bands. Oak, ash, elm or beech were the favourite woods.

The upper part of the trawl net, known as the 'back' was attached to the beam, but it obviously would not have done to drag the beam along the sea bed on top of the foot rope, or it would just snag on the first obstruction. Consequently the beam was fitted at each end with wrought iron trawl heads in an elongated D shape, forming a sort of toboggan, 3 feet to 4 feet high to lift up the beam and net clear of the ground. A pair of irons weighed from 230 to 360 lbs.

This heavy and complex fishing device had to be towed with the tide, and faster than the tide to extend the mouth of the net properly. Smacks with a trawl down needed to maintain a minimum speed of about two knots for some six hours. This was equivalent to a sailing speed of six to eight knots without a trawl.

To achieve this required a potentially fast sailing vessel and one which was heavily enough constructed to stand the stresses of towing, and absorb the strain of the wind which would otherwise drive her very much faster if she were not towing. The sort of wind which would produce a gross speed of seven or eight knots means a wind speed of 20 knots or so. This is about Force 6 in the Beaufort Scale, sometimes called a yachtsman's gale. It is not an unfair description because for a small cruising craft it is quite heavy weather. To the sailing trawler it was just favourable working conditions.

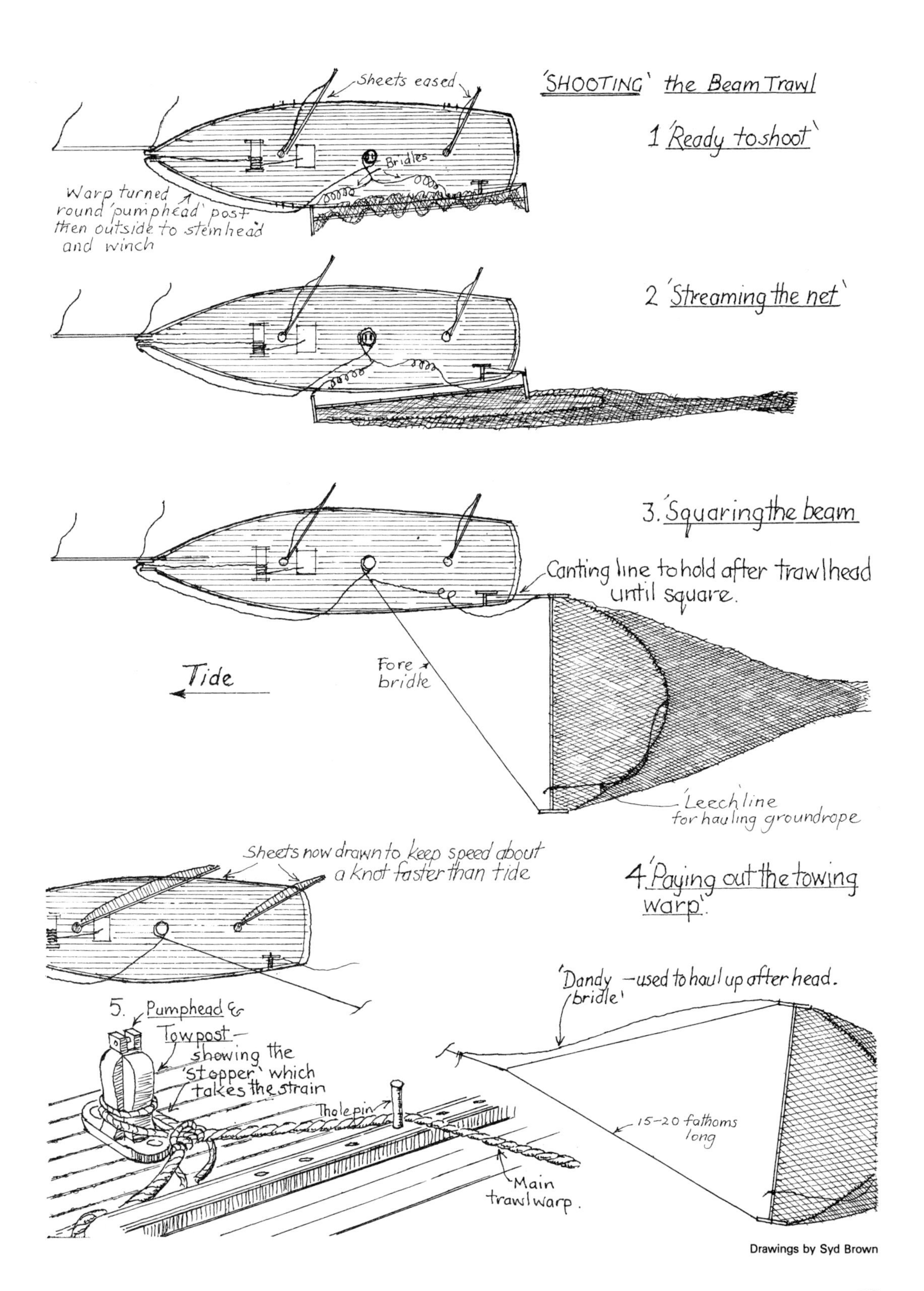

Drawings by Syd Brown

● A rare photograph of work at sea in sailing trawler days. It shows the belly of the trawl being hauled aboard *Torbay Lass,* over the beam of the trawl which is resting along the ship's rail. It was unearthed by Mr. Larry Bubeer, a sailing man of Brixham. *Torbay Lass* is still afloat as a yacht, re-named *Kenya Jacaranda.*

Drawings by Syd Brown

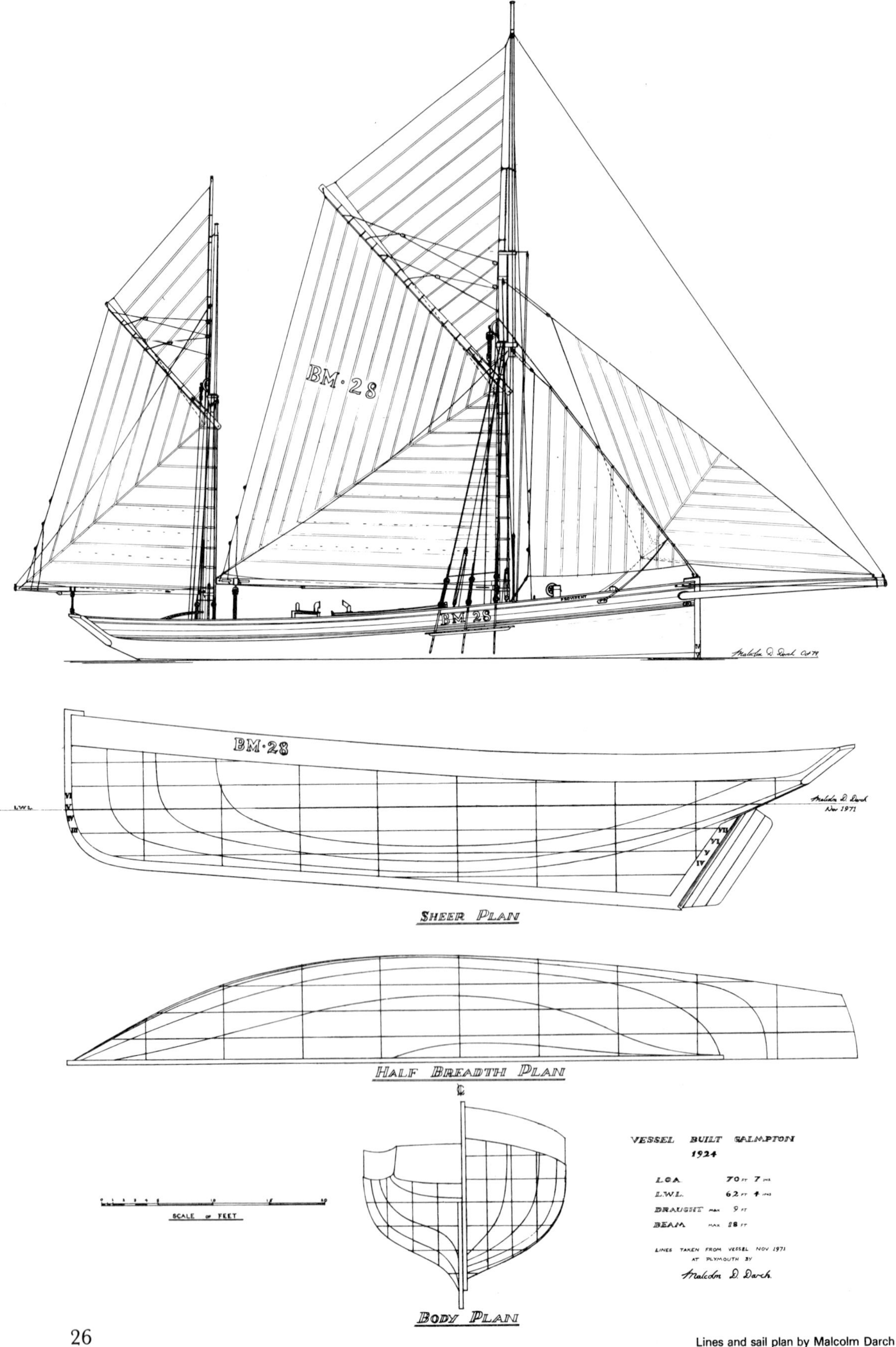

Lines and sail plan by Malcolm Darch

Evolution Of A Design

"THERE WAS nothing soft about a Brixham trawler . . . more than 3,000 years of trial and error, of triumph and disaster, of courage and endurance went into her construction. The Brixham trawler may have passed into the limbo of ancient and forgotten things but her creation was not in vain", wrote T. C. Lethbridge in "Boats and Boatmen".

It is true in a general sense that the design of a Brixham trawler stems from the tradition of ship building in Western Europe and the Mediterranean, which stretches back into classical times. It is founded on everything which has gone before, but its particular history covers only some 200 years. Any description of the first part of that period must be mainly based on certain suppositions. The second part presents much more hard evidence and, for the last 100 years or so, the record of the camera.

The design which has been evolved for the Brixham trawler, and which we see in its final form in *Provident*, is the result of what one may call a sort of 3:3:3 formula. There are three distinct elements in the design, namely, the hull below the waterline, the hull above the waterline, and the power plant. The last means the sail plan, the aerodynamics of the sails and the arrangement of the masts, spars and standing and running rigging.

These three elements are moulded by three sets of interlocking factors, hydro-dynamics in relation to the hull, aerodynamics in relation to the sail plan, and the peculiar demands of towing and working a trawl net. In the last must be included the human element of the trawlermen, but it can be safely advanced as an opinion that the comfort and convenience of the crew of a vessel have never been uppermost in the minds of designers. This applies as much to the most modern vessel afloat as to the Roman galley. The accommodation aft for the crew in a modern tanker may be extremely plush, but the noise and vibration from the engines is considerable and the steering position aft is in the worst possible place. A Brixham trawler's accommodation was not very grand, but at least it was in the part of the hull where the motion was easiest, and the steering position was in the right place.

The third 'three' in the formula lies in the mind of the designer. Today naval architecture is a formal profession based on academic study and designs are developed with the aid of applied mathematics, tank and wind tunnel tests. The designer of a traditional vessel had no such aids, except a rather limited application of mathematics. He proceeded on three bases—trial and error, inspiration and art imitating nature. Trial and error started with the first primitive man falling off a floating log. Inspiration came with the first refinement of a floating log, and remains essential to the modern naval architect, although the cynical might not see much evidence of it. Art imitating nature as a factor has been mainly superseded since the classic English concept of a hull form, resembling a cod's head and mackerel tail, was abandoned. Although it could re-appear; for instance the fact that dolphins can swim faster than their length defines as scientifically possible in relation to known factors, has attracted some attention.

In the past, also, development has always been slow. New inspirations have rarely received immediate acceptance. The seaman is conservative in outlook. The old design may have unsatisfactory features but the new one, for all anyone knows, may drown you.

The evidence for the early evolution of the Brixham trawler is thin and sketchy. The local word for the vessel is 'smack', which is a Dutch word. It would be fanciful to suggest that perhaps one of Dutch William's retinue landing at Brixham with the new king in 1688 may have stayed on to have some influence in the local scene, but there is no record.

Mr. John Horsley, former Curator of Brixham Museum, has assembled what evidence there is of early trawling activities in Torbay. In 1525 Leland reported objects being brought up from the bed of Torbay in nets and in the next century fresh soles were available at Exeter market. Both circumstances suggest trawling. A rough sketch exists showing a small vessel with square sail, which would have been capable of towing a trawl with a 15 foot beam. By about 1700 the square sail may have been replaced by the lug sail, with open, or part decked, boats, 25-30 feet in length. Fifty years later the craft was probably larger, fully decked and gaff-rigged. They were the prototype of all the trawlers which have come since then.

The eventual standard Brixham trawler hull, below the waterline, has two simply defined characteristics, which can be studied in *Provident's* lines, and in the photograph of her dried out, alongside the quay in Salcombe. They are a deep heel and a long keel. This means that looked at longitudinally the underwater hull has a relatively small draught at the bow and much more, nine feet in *Provident's* case, at the sternpost. The length of the keel is, roughly speaking, the same as the waterline length. This, with a straight stem, is a standard shape for traditional working craft in the British Isles. A modern yacht, from which a different kind of performance is demanded, has a keel length much shorter than that of the waterline.

The long keel and deep heel gives an easy motion in a seaway and good steering characteristics. Falmouth oyster dredgers, for instance, can be made to steer themselves quite happily. The easy motion is essential. Few modern yachtsmen practically live at sea, as trawlermen did, and most of them endure rough motion for relatively short periods. Trawlers had to have a good platform for hauling a trawl on board. The easiness of the motion in *Provident* is quite remarkable. Down below in heavy weather old hands can sleep undisturbed by any excessive pitching and rolling.

A wide deck was needed for hauling the trawl but too great a beam would, in fact, give too lively a motion. Too much sharpness in the bows would produce too wet a boat, but too much fullness would bring violent pitching and undue strain on gear and men. The hull shape is a compromise therefore and traditionally such compromises were arrived at in design by the construction of a half model. Given this model the prospective owner and skipper could approve or modify the design. Whatever shape they arrived at they were no doubt exposed, in a small maritime community, to quite unsolicited advice, headshakings and the mutterings in beards of ancient mariners.

● This photograph gives a clear idea of the lines of a sailing trawler with the long keel and deep heel in the main tradition of English inshore craft. The trawler stern and the comparatively low freeboard down aft are emphasised. She is dried out alongside Custom House Quay, Salcombe. In the foreground is a heavy clinker built West Country punt. It does not belong to *Provident*, but she would have carried a similar, slightly less chunky boat, on deck on the starboard side, in her working days. *Provident's* dimensions are 70 feet 7 ins length overall, 62 feet 4 ins length on waterline, 18 feet breadth, 9 feet maximum draught. It has been suggested that she was originally intended to be longer. When Malcolm Darch took off the lines he formed the opinion that there is apparent discontinuity in her lines towards the bow. Certainly *Provident's* shape is not as good as that of *Rulewater,* another Brixham smack which the Island Cruising Club also operated in the 1950s. It could be that *Provident* was shortened in response to economic and crewing problems at the time. Whatever happened, Sanders & Co. completed *Provident* the second in 1924 and she was launched from the same yard as her predecessor by a niece of William Pillar, a Miss Winifred Friend.

Author's photo.

Above the waterline the shape of a trawler is unlike that of most other fishing vessels. The bow is high, and the deck slopes up towards it. This characteristic can be seen in modern side trawlers and in most vessels of traditional design which have to face particularly heavy seas. The trawler sheer is particularly pronounced as about threequarters of the way aft the side of the vessel becomes low by contrast. There is quite a small 'freeboard'. Indeed with only a slight roll in fine weather *Provident* is capable of slopping water over a dry deck, to the discomfort of those who happen to be sunning themselves.

The reason for the low freeboard is the necessity of getting the beam trawl in and out over the bulwarks. The beam is heaved in by the capstan, laid on the rail, and after that the crew have to haul in the belly of the trawl.

Lastly the stern of a sailing trawler is either rounded or square, and gives a great deal of buoyancy. This is because when trawling the drag of the net causes her to squat heavily by the stern. A sharp stern, as in many sailing drifters, would not do.

Although initially many trawlers which fished in the North Sea were built in Brixham, the East Coast boats eventually evolved a different hull form. It had a shorter bow and was adapted for the short steep seas of the North Sea, while the Brixham trawler was more suited for the Atlantic rollers encountered off the South-West peninsula.

The evolution of the trawler's sail plan was equally involved and took place over many centuries. In medieval times and even as late as the 17th century the simple single square sail was retained by some coastal craft. In deep sea vessels the square sail was developed into a rig of great power, and complexity of rigging, by a process of multiplication. A single square sail for a relatively small craft proved inefficient. In Brixham the square sail was probably replaced in the 18th century by the lug sail, which is a square sail turned fore-and-aft. In effect it takes on some of the characteristics of a fore-and-aft sail while retaining some of its original ones. West Country drifters adopted the lug and retained it to the end of the sail era, as did the Scottish fishermen. The East Anglian ports went over to a ketch rig, but still continued to refer to their boats as luggers.

As trawlers grew larger the lug sail was not powerful enough and by the middle of the 18th century was being replaced by the sloop rig with fore-and-aft sails on a single mast. By 1785 the smacks were up to 40 feet long and two headsails were carried. By 1800 there was a topsail and probably a fidded topmast in place of the single pole mast. The rig would more properly be referred to today as a cutter.

In the middle of the century, with the stimulus of the North Sea adventures, the smacks reached 60-65 feet in length. To the standard headsails, consisting of staysail and jib was added the 'towing foresail'. This was a very large triangular sail set in place of the staysail and overlapping well aft of the mainmast. The clew and sheet were about two-thirds of the way aft. It was the working fore-runner of the yachtsman's 'Genoa' staysail. Used with the wind abeam it gave greatly increased power for towing the trawl.

Then about a hundred years ago the growing size of trawlers, with huge sail areas and main booms extending over the stern, led naturally to a division of the sail plan. The mainsail was reduced in size and a mizzen mast with gaff sail and a topsail set on a yard was added. It made a much more manageable rig for craft which had now grown to 70 feet in length.

The Brixham trawler had now reached its final form and there were no radical changes in hull and rig even when the last generation were being built in the 1920s. Some older vessels kept the original cutter rig, although they were still called 'sloops', until the end of their working lives. The newer and bigger ketches, with their mizzen masts, also continued to be called 'sloops', rather confusingly. They were also divided into two classes, those under 40 tons being referred to as 'mules' and those over as 'the big sloops'. The vessels which retained the cutter rig were also known as 'Mumble bees', although the connection with the Bristol Channel port is not too obvious. The nomenclature is confusing to those familiar with the generally accepted terms for various rigs but was perfectly clear to the fishing community!

The length and power of the trawler's hull, combined with its very large sail plan, made it capable of say seven knots, in what a smacksman would regard as a good breeze, without the trawl down. With the trawl down it would tow at say two knots. On passage, with a fair wind of near gale force a trawler could average over nine knots. One of the fastest, *Ibex,* once sailed from Trevose Head 140 miles to Plymouth in 15 hours. In September 1975 *Provident* herself, with a beam wind of Force 7, gusting to 8, covered nine and a quarter miles in one hour. What the old sailing trawlermen would have been able to get out of her in such conditions is anybody's guess. Faster times for sailing trawlers are claimed and still argued over in their home ports. For instance in the 1927 Brixham Smack Race the winner covered the course at 13.9 knots average, so must have been doing considerably more in the gusts. Most famous of all the Brixham trawlers was *Ibex* owned by the Upham family. Skipper Widger sailed her for 18 years from 1899 and she won the smack race so often that he was eventually asked not to enter her any more.

– *Glory BM16* built at Galmpton in 1906, owned by Samuel Pine Ellis and astern *Lily Doreen BM115* built at Brixham in 1921, owner J. P. Holland.
Photo: *The Times* Picture Library

Above & Below

IN RESPECT of her rig, sail plan, standing and running rigging *Provident* has remained virtually unaltered throughout her career, as far as general appearance is concerned. Inevitably in sailing so many thousands of miles, the equivalent of three times round the world between 1951 and 1976, there were accidents and spars had to be replaced. Sails also wore out, although some lasted a remarkably long time. An American cotton mainsail acquired in the 1930s lasted until the late 1950s. Her original towing staysail survived until about the same time. It was a great weight of canvas and took several people to carry it along the deck. The hanks which set it on the forestay were formidable pieces of ironmongery, vicious pointed spiral devices of iron which were twisted round the stay. At least there was nothing to go wrong with them, unlike modern spring piston hanks.

The purchases for setting up the peak and throat halyards were removed by the Island Cruising Club, as being unnecessary with a large crew available.

The main modifications came in the deck gear when *Provident* was converted to a yacht. The tiller was replaced by a wheel, using a powerful worm and pinion arrangement. Even then the steering could be tiring enough in heavy weather, but at least it was not capable of knocking people over like the old heavy tiller.

The steam capstan, which hauled in the trawl, was removed, together with the winch and windlass up forward, the pumphead amidships and the dandy wink, down aft on the port side.

The detailed procedure for shooting and hauling the beam trawl was fairly complex, but reduced to its essentials, it meant getting the heavy beam and net overboard and trailing it astern so that it sank to the bottom without mishap, eventually hauling it back up to the surface, getting the beam on the rail and finally lifting up the cod end, with perhaps three-quarters of a ton of fish in it.

The trawl was towed with 120 fathoms of six inch circumference, left hand laid, tarred rope, which was kept hard. This was led up from the warp room and around the steam capstan on the port side. Prior to the introduction of this equipment, which was first recorded in 1886, it would have gone over the windlass, forward of the mainmast. The warp then led over the roller, which can be seen on the port side of the stemhead. The end was brought aft, outside the rigging, and shackled to the main bridle on the trawl beam.

The after, or dandy bridle, was now run off the dandy wink, or winch, which is down aft on the port side. This bridle was led outside the mizzen rigging to be hitched on to the warp, just above the forward bridle shackle. A canting line was also rove through the after trawl head. Following this the net was shot overboard to hang from the beam. The fore bridle was then slacked away and the pressure of water would force the beam square to the stern. The canting line was then let go and the trawl would disappear under water. Increased way was put on the vessel to flatten the angle of the trawl warp. If the trawl was allowed to go straight down it might twist on its back.

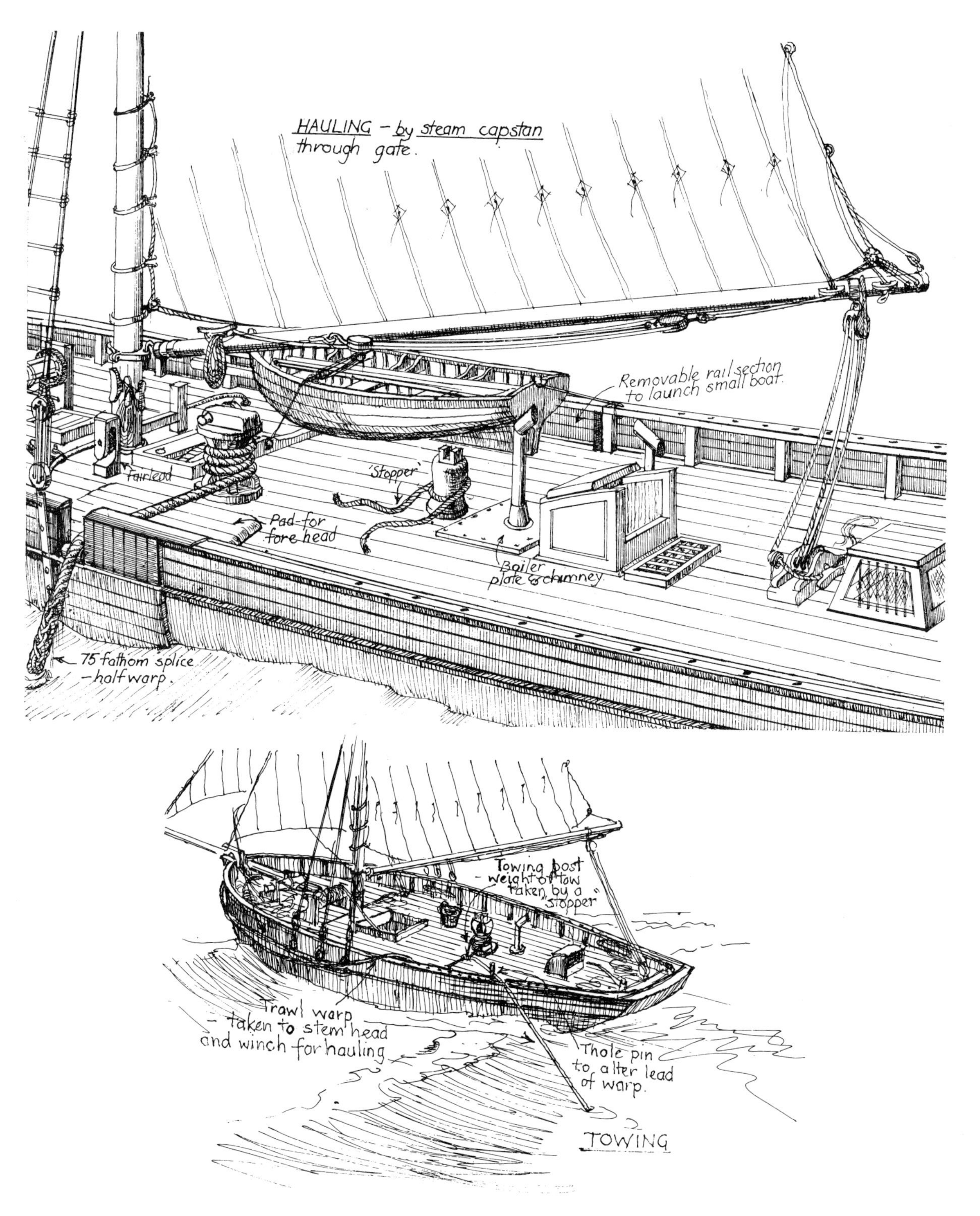

Drawings by Syd Brown

The warp was paid out according to the weather, tide, depth of water and nature of the ground. More wind meant more warp. The warp was led inside a big thole pin on the rail and secured to the pumphead by a 'stopper'. Two turns of the stopper were taken round the pumphead and it was hitched to the warp using both ends in overlapping turns. The stopper had a lesser breaking strain than the trawl warp so that should the trawl become foul on the bottom it would part. The slack of the warp would then go overboard without itself parting, and the smack would come up into the wind.

When the trawl came to be hauled the steam capstan made relatively light work of the process, compared with the old hand windlass which might take one to three hours for the job, according to wind and sea. On the beam coming to the surface the fore end was hauled up by a tackle or the fore halyard and the after end by the dandy wink. The beam was made fast and the belly of the net hauled in by hand. When the cod end containing the catch was reached a strop was passed around it and the fish tackle, on the fore side of the mizzen mast, used to hoist it up.

The normal crew for all this heavy work, not to mention the effort of sailing the smack, would be four men, or in *Provident's* time, only three men and a boy.

Down below *Provident* has been completely altered for her role as a yacht and virtually nothing remains of the original layout or the original fittings. However in Brixham in 1960 the late Fred Harris, who went to sea in *Provident* when she was brand new, was able to confirm that below decks she followed a standard pattern. Right forward, in the eyes of the ship, was the sail locker. Abaft that was space for the trawl warp and other warps, lines, blocks and bridles together with ship's gear of all sorts. The chain locker was also in this space just forward of the mainmast, keeping the weight out of the end of the ship.

Abaft the warp space came the fish room, with a locker for a ton of ice, fish pounds and two water tanks in the after wings. Then came the first bulkhead which sealed off the rather cramped boiler room cum galley. If the cook bent too far over his hot stove his after end came into painful contact with the boiler supplying steam for the capstan. For quick boiling of water the cook placed little iron saucepans, known as 'shoes', in the boiler fire. The boiler room also had to provide space for a coal bunker, more water tanks, lamp locker, and the cook's stores. Refrigeration was provided by a little hatch, known as a 'Norwester', through which food, which had to be kept cool, was pushed, to sit on top of one of the water tanks, in the lower temperature of the fish room.

- These two photographs are the only known surviving record of *Provident* in her fishing condition, half a century ago. Mr. Tony Pawlyn, a researcher on Newlyn, found them in the City of Bristol Museum, in albums which were the work of a Clevedon man named Keen. The weather appears to be drizzly, so unfortunately the photographs have no sharpness of detail, but we can see that there are no bobstay or bowsprit shrouds rigged, as there are today. There seems to be just enough wind to berth her alongside the quay, with some pulling and pushing to get her round the knuckle!

Photos: City of Bristol Museum.

BM·28

Next to the galley came the cabin with bunks for four men, although sleep was a comparatively rare commodity. There was a horse shoe shaped locker seat running around below the level of the bunks, a table, a stove, and some lockers for clothes and crockery. Abaft the mizzen mast there was a space for ship's gear. The smacks were innocent of such refinements as wash places or water closets. Fishermen belonged to what is known to yachtsmen as the 'bucket and chuck it' school. At sea fresh water is a scarce commodity in any small craft. The processes of ordinary cooking and brewing tea tend to use up most of the available supply and washing, especially washing clothes, is at a severe discount. One can be sure that the sailing smacksmen did not bother too much about them either.

During the 1920s *Provident* followed the Brixham man's seasonal fishing routine. In the winter they fished out of Brixham and at the end of March there was a move to Mount's Bay, but if the wind was westerly they tended to run back home to land the catch. If however the wind set in easterly, as it tends to in March, they landed in Newlyn. From Mount's Bay they progressed to the Seven Stones, off the Isles of Scilly, and by May the Brixham men were on the Tenby Grounds and landing fish in Swansea. There they stayed until August when they went back to Brixham for the Regatta.

By the end of the 1920s this life came to an end for *Provident*. Like most of her surviving sisters she was put up for sale and bought by a discerning American, Mr. R. Howe Lagarde, for conversion to a yacht. Remarkably, he did not install an engine.

● For many years through the fishing sail era men of the various ports took time off from fishing to compete in an annual local Smack Race. At Brixham, during Regatta Week, they used to race for the King George V Cup. The practice begun as long ago as the 1850s continued many years after the smacks had ceased fishing when the remaining examples of these splendid vessels were all sailing as yachts. Below opposite *Provident* is seen competing in the last Smack Race in 1953 when her only rivals were *Rulewater* and *Terminist.* Above, the start of a Smack Race in the 1920's. *Servabo BM77* (still afloat in the West Indies) is just ahead of *Lily Doreen BM115.* Unlike yachts the smacks raced from a standing start, the augmented crews of a dozen men having to jump from the punt alongside, weigh anchor and make sail. Faintly in the background can be seen the big racing yachts of the period assembled for Regatta Week.

Photos: Alfred Walker, Brixham
The Times Picture Library

Stars And Stripes And Green Burgee

FOR FIVE YEARS, from January 1933, *Provident* wore the Stars and Stripes. Her new owner had her converted at considerable expense by the famous British yacht designer Morgan Giles.

Working craft have always been attractive for conversion to yachts. Their sea-keeping qualities and their comfort, in terms of easy motion, deck space and space below have great appeal. In more recent years the vintage element in their appearance and character has added a special interest for owners with a feeling for tradition.

Against this they tend to be comparatively poor performers to windward. As the old sailing barge skipper said, "If I were a gennelman zur, I'd never go to wind'ard". It can be slow progress in a traditional vessel. Add to this the heaviness of the gear and the reliance on the muscle power that is sardonically called by seamen 'Armstrong's Patent', together with very heavy maintenance costs in a hull of formidable size and we have the other side of the coin.

Few can afford to maintain such craft, so while the smaller classes of ex-work boat remain popular, Falmouth quay punts, former luggers and so forth, the big trawlers and pilot cutters have dwindled over the years to a handful. Of the Brixham smacks, besides *Provident*, only *Vigilance BM76, Terminist BM321, Torbay Lass (Kenya Jacaranda)* and *Servabo* have survived in this country.

When *Provident* sailed from Brixham under the American flag she was probably never expected to be seen in her home port again. The new owner was quite a young man who had the money and the leisure to take up globe trotting as skipper of an ex-trawler as a way of life. The year 1935 found him cruising in the Mediterranean and at Malta the mate went sick. Lagarde signed on in his place a young Englishman called Ronald Powell. Powell had made himself a competent sailor and escaped to sea from a dreary shore job. To him we are indebted for a glimpse of *Provident's* life at this period, for in 1936 Jarrold's published his very readable account of his experiences under the title *I Sailed In the Morning.*

- View from the business end of *Provident's* bowsprit with a modern crew in a relaxed mood, cruising towards Belle Ile in the Bay of Biscay. Their ages range from twentyish to a veteran of 75 with a taste for active holidays. As a reminder of working days the sheave for the trawl warp can be seen on the port side of the stemhead. She is sailing in light airs and some unorthodox canvas has been set.

Author's photo.

According to Powell, Lagarde was a fine athlete and very strong man of about 170 lbs. (even then no description of an American was complete without his weight being stated.) He seemed to pursue, however, a rather solitary life as a sea-rover. Powell and the crew were confined to the focsle, as normal for paid yacht hands in those days. Lagarde occupied the big double stateroom aft and the rest of the ship including the saloon. In latter days the Island Cruising Club have packed 16 people and an engine into the same hull, in considerable comfort by yacht standards.

A winding staircase led to the glass panelled teak deckhouse over the companion. Photographs in Powell's book show that this deckhouse or doghouse was rectangular and narrower than the one which replaced it after the war. It is difficult to reconcile this with the story that the American owner asked for the deckhouse to be made to look like his Chrysler 'Airflow' saloon. The later doghouse has a sloping front and is generally considered not to accord very well with *Provident's* appearance. In fact it bears a considerable resemblance in style to the forward companionway hatch on *Cutty Sark.*

Down below the only reminder of fishing days was a 300 lb icebox for even in those days no American seemed able to contemplate life without ice.

On deck Lagarde made no changes of note in the rigging. Vangs for the peak of the gaff led to what could have been a serious accident and were discarded. He appears to have rigged running backstays, presumably to counter the effect of boomed out staysails when running in the Trade Winds.

To serve in *Provident,* according to Powell, was typical of American sailing ships. The skipper was rather a 'hard case', but the food was good and ample. Under British ownership the skipper might have been softer, but the food would have been miserable. Lagarde was a strict disciplinarian and drove the ship hard. In 1935 *Provident* ran from Palma to Europa Point, a distance of some 500 miles in 56 hours in a heavy north-easterly gale.

On the completion of his westbound crossing of the Atlantic in 1935 Lagarde suddenly abandoned his solitary sea life to get married and took his bride honeymooning in his ex-Brixham smack. The following year they crossed the Atlantic eastbound and encountered heavy weather to the Azores, being hove-to on two occasions. Her American owner gave *Provident* a new cotton mainsail of fine quality, but he recorded that American sailmakers would not tackle a loose-footed sail and that he found one laced to the boom not as good in heavy weather.

In November 1937 *Provident* passed to the ownership of Richard Finch Winfrey, a Lincolnshire farmer who settled at Helford with his family. During the Second World War *Provident* was laid up and escaped being used as a barrage balloon vessel, a fate which led ultimately to the demise of a number of historic sailing craft.

● The mainmast, opposite, is a single solid spar rising 40 feet above the deck, with the topmast above it making a total height to the truck of 68 feet.

Author's picture

- This beam view of *Provident* cruising up The Bag at Salcombe with the evening sunlight behind her, gives a striking impression of the lofty gaff rig of the sailing trawlers. The light shines through her modern terylene tan coloured sails as it would never have done through her original canvas, which was treated with a home brewed preservative mixture of grease, red ochre and seawater, known as 'cutch'. It is usually thought that gaff rig succeeded the old lug rig for fishing boats in pursuit of greater power for trawling, but ease of handling may have been a more important factor. Cornish fishermen, with their drifters, stayed with the luggers to the end of sail. In their case the number of men required to haul a fleet of drift nets, six or seven, corresponded with the number required to dip the big lugsail with its huge yard when going about. Perhaps surprisingly, a trawl required fewer men to haul it. It was therefore logical to adopt a rig which, although still heavy and with more rigging, required less men.

Photo: Ken Fraser.

After the war *Provident* was sold to a Mr. Fisher who kept her for a time in the dock at Penzance. From him she passed in 1949 to Brigadier T. E. N. Helby, M.C., who cruised her in the Baltic in the following year, not without some alarms and excursions and the demise of the 'Atlantic' petrol-paraffin engine, which had been fitted since Lagarde's day. This auxiliary was installed in the space aft in which Lagarde had kept his solitary state when owner/skipper.

The M.A.N. diesel which was then installed expired a few years later and was replaced by a 54 horsepower Lister, of generous proportions which proceeded to go on for ever.

In 1951 *Provident* became the flagship of the Island Cruising Club, founded at Salcombe by John Baylay the same year. It was a very fortunate event in every way. The type of yacht which *Provident* represented was of limited appeal. The enthusiasts to whom such a vessel did appeal were likely to lack either the necessary resources, time, expertise or crews with which to maintain and operate her. All too many yachts of this kind were entering a sad decline, for private owners with the leisure to use them and the money to maintain them were a dwindling breed.

The Island Cruising Club gave the survivor from the past a very long new lease of active life. The Club created 15 berths, later 16, in her and even then the accommodation remained spacious, although the arrangement might not have appealed to Mr. Lagarde.

With a full-time skipper, mate and cook appointed for summer cruising the Club has maintained and operated *Provident* with the income derived from the sailing charges made to members. The normal cruising ground has been the Brittany, Devon and Cornwall coasts to Ushant and Scilly and sometimes down into the Bay of Biscay; for special events, such as the Tall Ships Race, as far as Spain.

The distance covered in a season under the Club's green burgee would normally be in excess of 3,000 miles in a series of one week and two week cruises. Inevitably in such an active life, far in excess of that of an average yacht, there have been alarms and excursions from time to time and occasionally a broken spar. The Brittany coast has many hazards of rocks, narrow channels and strong tidal streams, although generally speaking they are lit and marked to a high standard. A Cornish yacht hand was once heard to say of it, "'Tis a wicked coast down along – a pile of rocks thicks as peas in a saucepan. I bear to mind faither allus said so. 'Tis the Lord's punishment. Wicked ways do surely bring a wicked coast. I do reckon you'm better in our little ol' waters. Can't equal 'em".

However this may be, the Island Cruising Club has never been deterred from exploring every haven of the Brittany coast, generally without any major incident. Her narrowest escape was leaving the entrance to the Treguier River on one occasion when a hail squall suddenly reduced visibility effectively to nil and somehow she got out of the proper channel and went gently aground on a patch of sand among formidable rocks. In due course French bureaucracy descended on the scene with remarkable efficiency. A suitable plan was drawn up, local pilots provided, and when the tide served *Provident* threaded her way out, safe and sound.

Over the years the old trawler has been cared for devotedly by a succession of Club skippers, mates, cooks and crews, and by the Club as a whole. In 1971 she was purchased from the Club for the Maritime Trust through the generosity of Mr. Philip Pensabene and Philip & Sons Shipyard of Dartmouth. The Trust promptly handed her back on charter to the Club, retaining responsibility for maintenance work towards her preservation.

So *Provident* has continued at sea every summer under the burgee of the Club and flying the flag of the Maritime Trust from the spreaders. Members continued to regard her as the Club's flagship. Perhaps not all of them would go so far as to say, as some do, that for them '*Provident* is the Club'. But the fact that some of them do say so is an indication of what she has meant to them over the years. The Club's input in respect of *Provident* has been enormous over the period 1951 to date, in terms of time, money, labour and devoted care and thought for her maintenance. In return her output has been vast. If that vogue word charisma can be applied to a ship it can be applied to *Provident*. To devotees she has a detectable air of benevolence and has been a focus of happiness for thousands of amateur sailors, and not a few professionals.

Her later career has been something which William Pillar could scarcely have foreseen when he had her built to replace the first *Provident*. That vessel was inevitably associated with war, death and destruction at sea. It is fitting that the second *Provident* should have had such a peaceful and happy role to play.

● Opposite, the starboard main shrouds on *Provident*. The steel shrouds are siezed around wooden deadeyes and made fast to the deadeyes on the chain plates by lanyards. This is the traditional method and has the merit of providing a degree of elasticity which responds to any jerking of the rigging from the vessel's motion. The shrouds are prevented from twisting by the wooden sheerpole above the deadeyes, which also acts as a pin rail. The forward pin has the staysail sheet belayed on it. The aftermost one is used for hanging up the heavy coil of the fall of the peak halliard, by means of a lanyard looped through it. The fall is actually belayed on a big steel pin on the kevil rail, below the ship's rail, after being led round a sheave under the kevil. Abaft the peak halliard is the single topping lift. The purchase hooked on to the forward part of the kevil is used to set up the jib, which is set flying. The massive and simple nature of the shrouds is a feature of gaff rig, which was evolved empirically and not engineered in the modern sense. Bermudan rig is finely engineered but the failure of one component can be disastrous.

Author's picture

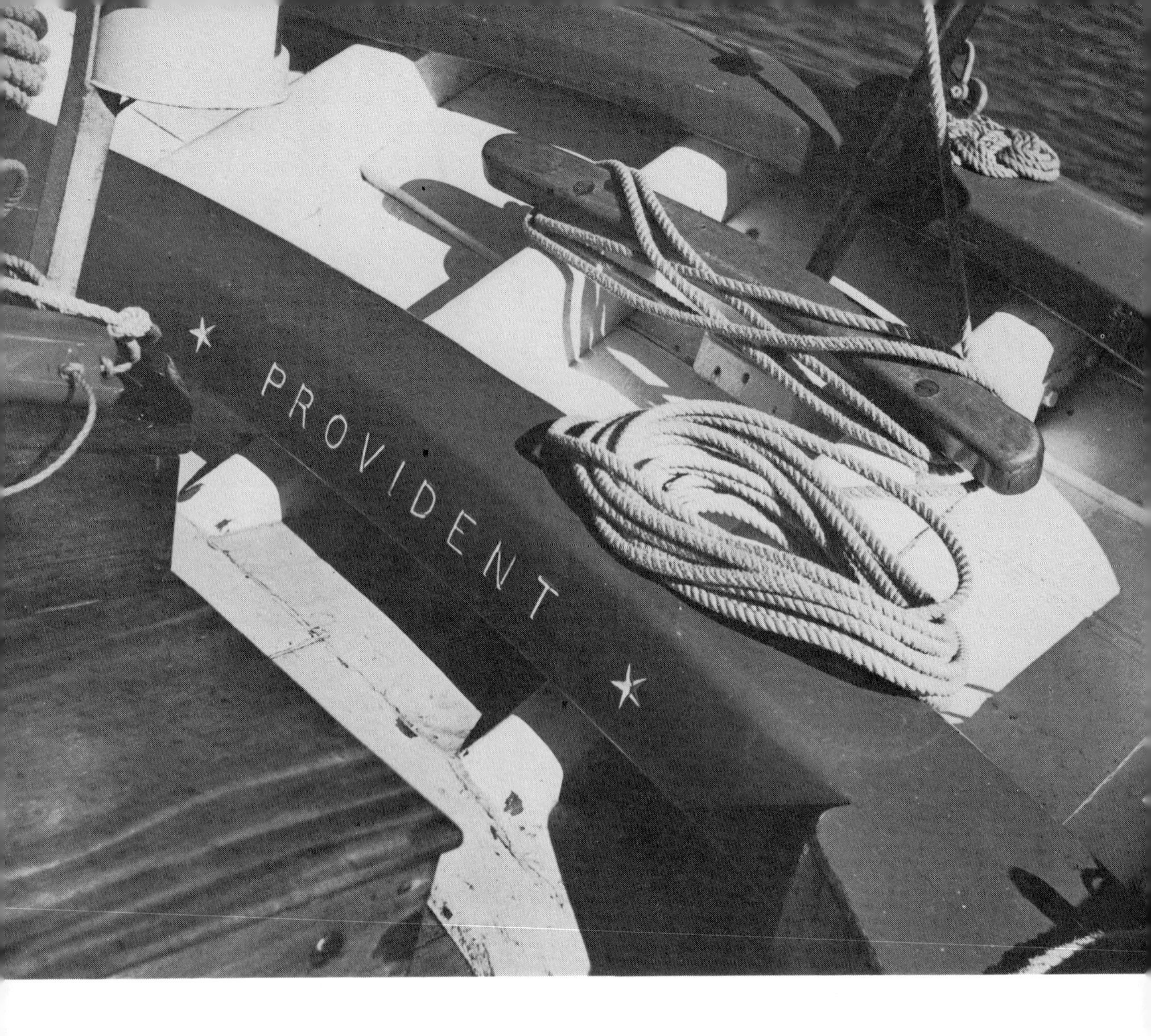

- Above, the transit rail, with the vessel's name picked out, stern timbers and taffrail. In between is the cleat for the mizzen sheet. This photograph, taken at the end of the season, is indicative of the high standard of maintenance on board *Provident* today. She would not have looked anything like as smart on her working days. The white haired man looking over the taffrail of *Provident,* with her Island Cruising Club Skipper, Chris Spencer Chapman, is George Bond. With four other shipwrights, Philip Wills, Tom Hind, Charlie Helley and George Moor, together with three of the Sanders family who owned the Galmpton yard, he worked on the building of *Provident* in 1924, which took six months. George Bond recalls that the timber was felled ten years before, and that it lay in the stack where it was felled until building commenced, when the brambles were cleared away and two horses, 'Blossom' and 'Prince' dragged the logs to the sawmill.

Photos: Tony Allen.
Author

PROVIDENT
BRIXHAM

Heavy Weather

AN OLD saying amongst sailors is "Grumble you may, but go you must". It is often true for the amateur as well as the professional.

Anyone who chooses a holiday which involves getting out of a warm bunk after midnight and putting to sea in a sailing trawler, with wind force 6-8 forecast, may be regarded as eccentric. Most of the crew of *Provident* one September night in Alderney Harbour would have admitted the judgement. She was due back on the other side of the Channel within 24 hours. The 1755 forecast gave a possible force 9 for Wight and Portland sea areas; the 0030 forecast mentioned nothing over force 8, while the reported wind at St. Helier, Jersey, was a mere force 5. The skipper decided to sail. Alderney Harbour is an uncomfortable anchorage in bad weather, but it promised to be much more uncomfortable outside.

For the other occupants of the anchorage wind and weather may have been enough to disturb their sleep, but *Provident* probably ensured their wakefulness. For any who peered grumpily out through their scuttles she would have presented a dramatic picture, preparing for sea, with all lights switched on, and the crew stumbling about the deck, in their oilskins, to the accompaniment of flapping canvas and creaking spars. To all this the rhythmic clank of the anchor cable being wound in formed a steady descant.

Leaving the harbour under sail involved a careful gybe and once outside, with a beam wind, it did not seem so bad after all, but a thunderstorm with a violent squall caused the ship to roll heavily, a reminder that the weather could be all the forecast said.

At 0600 the wind was still well on the beam and *Provident* was logging eight knots with full main and mizzen set, staysail and No. 3 jib. The last sail, and the absence of topsails were the only concession to the weather. Steering seemed not too difficult, but wind and sea began to increase. Every so often, perhaps as a wave gathered under the quarter and there was a surfing action, the ship would run straight and true on course with hardly a spoke needed on the wheel. Then she would fall off into a trough, gripe to windward and demand a considerable effort to get the wheel over to starboard and maintain the course. A three quarters of an hour 'trick' at the wheel was enough for the helmsman.

Between 0700 and 0800 some rain squalls came along, heavy enough to flatten off the seas. In that hour she logged nine and a quarter miles, so it is fair to assume that *Provident* must have been touching over 10 knots on occasions. In seven and a quarter hours just over 60 miles were covered on that passage. It was extremely wet on deck, but only once, when altering course for a steamer, did anything that could be described as a heavy sea come on board.

● A fresh breeze has water washing in the scuppers on board *Provident*. The towing staysail is set. Because of its power it has to be sheeted in with a 'watch tackle', a purchase consisting of a double and single block.
Photo: Mike Brownlee.

The Brixham trawler's mainsail was made with vertical cloths, the theory being that if a heavy sea burst the mainsail there would perhaps be a useable area left. The mind of the summer yachtsmen boggles at the thought of the violence of weather which could bring a sea on board to burst *Provident's* mainsail. Some years ago Mr. Howard Thomas a retired smacksman described to me a winter gale in the 1930s aboard the smack *Inspire BM13.* "We had our gear down when it came on to blow a very stiff breeze of wind (sic). So we handed the gear while already under two reefs in the mainsail. The skipper said that the glass was very low and we had better haul the jib in. The staysail was already reefed, and we put two reefs in the mizzen. It then came on very stiff indeed and we stowed the mizzen altogether. By this time we had the topmast housed too. So we ran for it."

Force 6, wind speed 22-27 knots, is defined in the Beaufort scale as a 'strong breeze'. Mr. Thomas's "very stiff breeze of wind" was probably force 8, a full gale, with winds of 34-40 knots. When it "came on very stiff indeed" it was probably force 10, a storm, 48-55 knots, or even worse. It certainly remained a lively memory for him 40 years later.

The kind of weather which yachtsmen meet in exceptional circumstances, like the trip back from Alderney described above, was routine in winter for the smacksman of old and he had to shoot and haul his gear in that sort of weather. Doing 10 or 11 knots in heavy weather in a vessel like *Provident*, just once in a while, can be exciting and exhilarating for amateur sailors. They know that they will not have to face it day after day in order to earn a living. The professionals endured life aboard such craft as *Provident* because they had been brought up to it; it was perhaps the only way of making a living, and a poor one at that. Whatever else it did, it made them superb seamen.

- Sea conditions like those above are an occasional thrill for today's yachtsmen crew, but were everyday working conditions for the fishermen. Left, *Guess Again BM346,* 24 tons, taken in 1926. She had earlier been the command of Skipper Widger, noted in the port for his racing prowess. She was locally built in 1918 and owned by A. Phillips of Brixham. Opposite, crew at work aboard *Torbay Lass* in harbour. They include the owner Alf Lovis, second from left, 'Johnner' Cole, extreme left and Bill Parnall, standing by the steam capstan.

Photos: Mike Brownlee
L. Bubeer

Shoreside Life

GOING to sea in *Provident* today gives the crew some impression, just a very small impression, of what life afloat was like for the sailing trawlermen. But it cannot adequately convey the feeling of hard days of unremitting toil and discomfort in all winds and weathers, winter and summer.

Their life ashore in Brixham, and the lives of their dependents, is possibly even more difficult for us now to envisage. We are separated from the Brixham of the sailing trawlers not only by time, half a century and more, but a great gulf in the style of material life.

A smacksman's pay was not likely to maintain him and his family in any great affluence ashore. The proceeds of landings were divided into seven shares, one and a quarter for the master, one for each man and three quarters for the ship. The boy only got 'stocker' money which was paid for out of items like fish livers and roes, or odds and ends of shipboard junk. If you caught nothing you earned nothing. Fred Harris who served in *Provident* in the 1920s remembered getting one shilling and sixpence (seven and a half new pence) for one weeks work. Normal earnings were then about ten shillings (50 pence) to 25 shillings (£1.25p a week).

On the whole, however, the share system was probably less of an evil in Brixham than in other ports on account of the fact that a higher proportion of the smacks were skipper or family owned. In Hull and Grimsby they were mostly owned by outside capital. This also meant that Brixham escaped the pernicious practice of 'fleeting' by which trawlers were kept at sea for many weeks on end, while carrier vessels were sent by the companies to bring their catches back to market.

If the fleet was kept in harbour during winter gales, there was no pay and no unemployment benefit. If the breadwinner was drowned or incapacitated by accident at sea there was no social security for the family. A measure of charity was distributed from time to time, but could not have gone very far.

Old photographs of Brixham show no quaint houses painted in attractive colours, but rather depressing looking dwellings, many of them occupied by

- Brixham trawlers often used Newlyn as a port of landing, sending their fish up country from the Great Western Railway's terminus at Penzance. Some fish was sold locally by retailers who worked direct from the foreshore where the fish was landed. This Newlyn woman is a fish 'jouster' or seller, who would have carried a wicker basket called a cowal on her shoulders supported by a band around her hat. She knits while she waits, a continual occupation between other tasks for the womenfolk of the fishing communities. The pony traps belong to male fish jousters who toured the streets of Penzance and hawked their wares with the aid of leathern lungs, which could produce a shattering bellow.

Photo: Alfred Robinson.

SAILMAKER

more than one family. The rent went either to a private landlord, or, if one family actually owned the property they would rent part of it to another, leading to considerable overcrowding. Six children might sleep in one room in one big bed.

A typical fisherman's kitchen cum living room can be seen reconstructed in Brixham's museum. The focus of the room was the kitchen range in which a coal fire provided heat for the room and for cooking and a cheerful glow. Gas became available in the town as early as 1842 but would only have been used for lighting in the more prosperous houses until a much later date. The poorer people relied on candles and, later, paraffin lamps. Late in the 19th century the typical Victorian tunnel back terrace houses appeared and were occupied by the more prosperous middle class, including trawler owners, who might provide accommodation for their apprentices. Such houses remain popular today, but as second homes transformed by modern decorations, electricity and a host of consumer durables.

In the past 'consumer durables' were very durable indeed and the modest equipment of a household was expected to last a lifetime, at least. The coppersmith would make a 'tin' kettle for one shilling and sixpence (7½p) while a copper kettle at ten shillings (50p) was a major investment. The tea which the kettle would brew cost two shillings (10p) a pound in 1883 and for two shillings and sixpence (12½p) you could get 17½ lbs of flour.

Not surprisingly fish formed a large element of the diet of Brixham people, but not, generally speaking, from the landings of the trawlers. Doubtless those on board took some fish home with them but the best of the catch went to market. The only part destined for local consumption, at any rate during the 1880s, was the 'stocker' which was regarded as unmarketable but edible.

The local drifters and inshore fishermen provided most of what was eaten locally, particularly mackerel. A watch was kept from Prospect House for the gulls which indicated the presence of a shoal and a flag run up to signal the drift net boats away to sea. Hake, ling, skate and whiting were also popular local fish, but to eat dogfish (although it can be very good eating) was to sink to the bottom of the social scale. When whiting were plentiful the women would preserve them with rock salt.

- Mumble Bees were a smaller class of cutter rigged vessels, perhaps named because they derived from inshore smacks purchased from Mumbles in the Bristol Channel. The one above opposite is the *Lizzie,* owner and master G. Friend. In the left foreground is the pumphead, to which was made fast a stopper from the trawl warp. The stopper had a lesser breaking strain so that if the trawl came foul on the bottom it would part. The loose warp would then fly overboard and the smack would come up into the wind, without parting the main gear. In the group are Percy Friend (sitting), Tom Gardner (right behind the others), Abe Bartlett, Dan Friend and Bill Lill. Below shows Brixham harbour early in this century. On the right *Boy John BM241,* 38 tons, built in 1905 and owned by J. G. Hill.

Photos: Larry Bubeer and F. S. Lyddon

Nineteenth century housing development made little provision for gardens, even in the better terrace houses. Moreover the configuration of Brixham, built on steep slopes around the harbour, put space at a premium. Vegetables and fruit would therefore have to be bought from local sources. Exotic fruit, like oranges, was probably confined to an annual treat in the Christmas stocking.

The elite of the shopkeepers were the butcher, the baker and the draper. The baker had the now forgotten function of letting space in his ovens during the day for families who had no proper cooking facilities at home.

Fishermen's jerseys would have formed no part of the draper's stock. They were knitted at home by the womenfolk, who were also employed in net-making, along with old or disabled fishermen.

Knitting in Brixham provides an unlikely piece of historical evidence concerning the colonisation of the Yorkshire coast by Brixham people. In the Museum are knitting sticks for a single needle, a method peculiar to Yorkshire. They were brought to Brixham in 1787, which suggests that the colonisation of the Yorkshire coast or at least regular fishing there by the Devon men may have been half a century earlier than had been supposed.

One of the few shops in the town selling non-essentials, if not the only one, was the sweet shop. It produced its own toffee and real bullseyes, the size of a Chelsea bun! For the youth of the town 'a penn'orth of sweets' would be one of the few paid for indulgences allowed them. Their life in Brixham was scarcely a privileged one. The 1870 Education Act brought order to their learning for the first time, but even in the middle of the next decade the Dame Schools lingered on. This was attributable to 'the old grey widow-maker' which brought poverty to the home of many a seaman's wife in the course of a year. It was accepted practice for them to set up a Dame School to provide a little income, regardless of any lack of teaching qualification. Indeed it mattered not if they were illiterate.

In 1850 Brixham had a free school on the national system for 130 boys, and 110 girls who paid one penny per week to attend. There was also an orphanage and a Reformatory in Berry Head Road. The Reformatory was a prime source of apprentices for the trawlers. Until 1883 an apprenticeship lasted seven years and was then reduced to five. This system, a common source of cheap labour, lasted until the 1930s and was seriously abused in the 19th century. Cases occurred of boys of 11 being bound apprentice for 10 years, for which there was no legal remedy. However, they generally fared better than in some other ports. The skipper of the Hull smack *Rising Sun* was hanged for murdering an apprentice. In 1883 the *Dartmouth and Brixham Chronicle* reported the murder of another apprentice in a Hull smack and a case of "cruel, debasing and disgusting" treatment of two lads at sea and called for legislation. Abuse of apprentices in Grimsby was rife and many got themselves imprisoned as a preferable alternative to serving any longer at sea with their masters.

The only regular forms of social gathering and entertainment were provided by going to church or chapel on the one hand or the pub on the other. When a lot of men were kept ashore by bad weather in the winter attempts were sometimes made by well meaning folk to provide them with 'healthy and instructive recreation' during the day in the Wesleyan School-room – or in other words to keep them out of the pubs.

Beer was cheap, mild and stupefying and until the First World War the pubs were open all day. There were also more of them per head of population. In Brixham, in 1850, with a population of 7,000, but with 1,600 seamen employed, there were no less than 20 inns and taverns, with six beerhouses in addition.

For the womenfolk the main relief from the round of cooking, cleaning, net-mending and knitting was probably the enjoyment of local gossip, but an occasional highlight was provided in the fetes and teas sometimes organised by individuals and institutions.

The weekly issues of the *Dartmouth and Brixham Chronicle* record many of them and throw some interesting sidelights on life in the heyday of Devon's chief fishing port.

On January 12th, 1883 it reported that Lady Churston had given a New Year's tea to the poor and also a hundredweight of coal to each of the poor attending a Mission service.

In the same week the weather stopped all fishing and kept 1,360 men and boys in port and the schooner *Freedom* went missing with her crew of four.

A week later Commander Phillpott, R.N. celebrated his wedding by entertaining 40 old people to a high tea, anyone over 55 being classified in those days as old. The weather had abated and 90 trawlers each landed between 30 and 75 baskets from which each boat earned between £3 and £14.

In June the Brixham branch of the Rational Sick and Burial Association livened things up a bit with a Grand Fete and Tea. By this time most of the fleet were fishing the North Sea grounds and earning around £20 per week. The Totnes Guardians were reported to have taken the momentous decision to halve the number of tramps seeking shelter in the workhouse by obliging them to have a cold bath.

Among such diverting news items the harsh realities of the sea are never absent for long, commonplace enough to excite no undue headlines. On June 8th the apprentice of the smack *Majestic* drowned when repairing the jib tack on the end of the bowsprit. His father and brother had also been drowned at sea. His widowed mother's plight was not uncommon. Others around her had given as many menfolk to the sea.

But already the railways were opening up this tiny community and almost imperceptibly standards of living began to rise. The railway not only took fish to market. Visitors came down the Great Iron Road to the West. After the coming of the internal combustion engine the flow became a flood. A new industry was being born. In time it would become more important than the fish.

Into this changing world *Provident* was launched. Few could have guessed that she would be one of the last of her breed. Nor for that matter would they have guessed that she would sail for so long, under such strange flags, and survive as a memorial to a vanished race of superb seamen and splendid ships.

- Above, ice carriers from Norway in Brixham harbour in 1868. The three masted topsail schooner is discharging her ice alongside the quay on the left, her ship's boat slung over the stern Baltic fashion. Before the advent of modern refrigeration methods ice was a valuable international cargo, with Norway a leading producer. Large blocks were cut with saws from the frozen lakes in winter and shipped packed in sawdust. Initially it was a hazardous cargo, as the interaction of the ice and sawdust produced an inflammable gas, leading to fires until the problem was recognised. When the railways opened up the inland markets for fish, the demand became enormous until ice making factories were set up at the fishing ports about a decade later. A Brixham smack going away to work the Irish grounds would take on board as much as two and a half tons. Outside the harbour a fleet of smacks can be seen waiting for a breeze. They are mostly cutters as the move over to ketch rig, as vessels become larger, had not yet gained momentum, but one ketch is visible. All the registration numbers are DH (for Dartmouth) because Brixham did not become the port of registry until 1902. By 1897 when the photograph at the top opposite was taken there were many more of the big ketches to be seen in the harbour and the number of smacks registered in Brixham was around 200. Although few fortunes were made in the fishing boom careful skipper owners could put by a nest egg, with luck, and build a good class of artisan house like those in the picture. By the time the bottom photograph opposite was taken in 1926 the heyday of fishing sail was over and the number of ketches in harbour had sadly dwindled. *'H.C.B.' BM373* with sails unbent looks rather neglected, but *Valerian BM161,* next but one behind her, built only three years previously, has her gear on board. Motorised invasion of the West Country had already by this time begun as the row of cars on the front shows.

Photos: D. F. & H. M. Stone, Brixham.

BRIXHAM INNER HARBOUR 1897

78490F. Brixham Harbour in 1926.

The Island Cruising Club

THE ISLAND Cruising Club which has had the care of *Provident* for the best part of 30 years, was founded in 1951 by John Baylay, at that time head of a Salcombe chandlery business. His objective was to give people who might not otherwise be able to do so, the opportunity to go cruising under sail and learn seamanship.

It was in many ways a pioneering venture. The boom in sailing as a recreation had hardly begun, there did not then exist the number of sail training organisations and other means of adventure holidays which there are now. At that time a yacht or sailing club invariably meant an association of people who all owned their own craft. The concept of a Club whose members owned and operated a fleet of vessels on a co-operative basis was an unusual one.

As well as *Provident* the I.C.C. acquired the elegant 1908 vintage schooner *Hoshi* (above opposite) and has down the years sailed a number of other traditional craft. The present fleet includes, however, smaller modern yachts as well. Club cruises have ranged as far as St. Kilda, remotest of the Outer Hebrides and as far south as Lisbon, but most cruises are around the English Channel and Brittany coast, with one or two a year to Ireland or Scotland.

There is also a thriving Club dinghy sailing section with floating headquarters and accommodation in *Egremont* (below opposite) an ex-Mersey ferry which is itself a preserved example of a disappearing type. Although the dinghy section has a large fleet of modern class dinghies and keel boats of many sorts, it manifests the same love of traditional sail in its ownership of several Salcombe yawls (notably a local racing class derived from a fishing boat design) and in a number of clinker built 13 foot dinghies specially made for Club instruction in the early years.

Training has always been an important part of I.C.C. activities. From an early date it established its own system of qualifications, but adopted the Royal Yachting Association scheme when this came in and it is a recognised R.Y.A. training establishment.

The Club now has about 3,000 members, consisting of people of all ages and from all walks of life and drawn from all over the British Isles and some from overseas. Members pay an annual subscription and charges are on a weekly basis when they go cruising or dinghy sailing.

The nucleus of people who administer and operate the Club and its fleet (a job calling for a considerable amount of dedication) are necessarily full-time staff members, but policy is decided by an annually elected Committee.

A famous Admiral once said that it took a comparatively short time to build a ship, but a long time to build a tradition. The Island Cruising Club has now been a long time building its own tradition and much of it is embodied in *Provident*, the flagship of the fleet.

Full details about membership of the Island Cruising Club can be obtained from The Secretary, I.C.C., Island Street, Salcombe, Devon.

The Maritime Trust

THE MARITIME TRUST was founded by HRH The Duke of Edinburgh in 1969, to rescue, restore and display vessels of technical or historical importance in Britain's maritime heritage, and to help and encourage others to do so.

The Trust's interest is not confined to any particular type of craft but each is judged on its place in our seafaring traditions. The Brixham trawler, described in this book, would be an obvious candidate in any list but in its first ten years the Trust has also acquired about 20 craft large and small, from a Falmouth oyster dredger to the hull of the 1860 iron 'battleship' *Warrior.* It has also helped numerous other ship preservation societies.

The Trust's vessels, which now include the *Cutty Sark,* are often the very last of their type and even with the greatest care can never again be seagoing; but in some cases, as with the *Provident,* the Trust aims to keep them sailing or steaming for as long as possible.

In 1979 The Maritime Trust concentrated a collection of its larger vessels in London, in the historic St. Katharine's Dock, near the Tower. This was done primarily to allow these vessels to be seen by a much greater number of visitors and to allow the Trust to build up a pool of skilled labour to keep the ships in good condition.

This Historic Ship Collection includes the West Country topsail trading schooner *Kathleen & May,* the Thames spritsail barge *Cambria,* the North Sea steam herring drifter *Lydia Eva* and the veteran steam coaster *Robin,* built in London in 1890. The steam tug *Challenge* and the Nore Light Vessel, on loan to the Trust, are also on show. The *Robin* and the *Lydia Eva,* which steamed under their own power to the Port of London, are kept in seagoing condition.

The Historic Ship Collection will not reduce the Trust's interest in smaller craft elsewhere, such as the Barry pilot cutter *Kindly Light,* now restored and on show at the Welsh Industrial and Maritime Museum at Cardiff. The fishing mule *Blossom* and the foyman's coble *Peggy,* typical small boats of the north east coast, are with Tyne & Wear Museums. The Cornish fishing boats – the oyster dredger *Softwings* and the St. Ives lugger *Barnabas* (right) – are kept in sailing condition.

Also in 1979 The Maritime Trust acquired HMS *Warrior* (1860) and the Royal Research Ship *Discovery,* two major vessels of great national importance. Work on them will take several years: the *Warrior* will then join other warships at Portsmouth, while RRS *Discovery* will remain in London and, in collaboration with the National Maritime Museum, will be opened to the public as a museum of exploration.

The Trust is a registered charity and its work of preserving and restoring these old ships – a costly task – is entirely dependent on public generosity. Further information can be obtained from the Director, The Maritime Trust, 16 Ebury Street, London SW1W 0LH.

SS 634
634
BARNABAS
SS 634